T0176347

GREAT VOICES IN RESEARCH
A CNRS EDITIONS / DE VIVE VOIX COLLECTION

This book was originally published in the collection Les Grandes
Voix de la Recherche, which presents the work of winners of the
CNRS Gold Medal, the highest scientific research award in France.
In short and lively texts, Gold Medal winners describe their pro-
fessional journey, share their passion, and reveal their work. Pre-
senting accessible, up-to-date content reflecting the most recent
advances in the sciences, they introduce the sine qua non in French
scientific research. As guides and mediators, these great voices
in research explore all areas of knowledge and, in clear, readable
language, present the major challenges facing the sciences today.

EINSTEIN AND THE QUANTUM R

EINSTEIN AND THE QUANTUM REVOLUTIONS

GREAT VOICES IN RESEARCH
A CNRS EDITIONS / DE VIVE VOIX COLLECTION

This book was originally published in the collection Les Grandes
Voix de la Recherche, which presents the work of winners of the
CNRS Gold Medal, the highest scientific research award in France.
In short and lively texts, Gold Medal winners describe their pro-
fessional journey, share their passion, and reveal their work. Pre-
senting accessible, up-to-date content reflecting the most recent
advances in the sciences, they introduce the sine qua non in French
scientific research. As guides and mediators, these great voices
in research explore all areas of knowledge and, in clear, readable
language, present the major challenges facing the sciences today.

EINSTEIN

and the

QUANTUM

REVOLUTIONS

Alain Aspect

FOREWORD BY DAVID KAISER

TRANSLATED FROM THE FRENCH BY
TERESA LAVENDER FAGAN

University of Chicago Press
Chicago and London

The University of Chicago Press, Chicago 60637
The University of Chicago Press, Ltd., London
© 2024 by The University of Chicago
All rights reserved. No part of this book may be used or reproduced in any
manner whatsoever without written permission, except in the case of brief
quotations in critical articles and reviews. For more information, contact the
University of Chicago Press, 1427 E. 60th St., Chicago, IL 60637.
Published 2024
Printed in Canada

33 32 31 30 29 28 27 26 25 24   1 2 3 4 5

ISBN-13: 978-0-226-83201-2 (cloth)
ISBN-13: 978-0-226-83202-9 (e-book)
DOI: https://doi.org/10.7208/chicago/9780226832029.001.0001

Originally published as *Einstein et les révolutions quantiques* by Alain Aspect ©
CNRS Éditions, Paris, 2019

The University of Chicago Press gratefully acknowledges the generous support of
the France Chicago Center toward the translation and publication of this book.

Library of Congress Cataloging-in-Publication Data

Names: Aspect, Alain, author. | Kaiser, David, writer of foreword. | Fagan, Teresa
Lavender, translator.
Title: Einstein and the quantum revolutions / Alain Aspect ; foreword by David
Kaiser ; translated from the French by Teresa Lavender Fagan.
Other titles: Einstein et les révolutions quantiques. English | France Chicago
collection.
Description: Chicago ; London : The University of Chicago Press, 2024. | "Great
voices in research : A CNRS Editions / De Vive Voix collection. This book
was originally published in the collection Les Grandes Voix de la Recherche,
which presents the work of winners of the CNRS Gold Medal, the highest
scientific research award in France."—Page facing title page.
Identifiers: LCCN 2024023272 | ISBN 9780226832012 (cloth) | ISBN 9780226832029
(e-book)
Subjects: LCSH: Einstein, Albert, 1879–1955. | Quantum theory. | Quantum
theory—History. | Quantum entanglement.
Classification: LCC QC173.98.A7613 2024 | DDC 530.12—dc23/eng/20240606
LC record available at https://lccn.loc.gov/2024023272s

♾ This paper meets the requirements of ANSI/NISO Z39.48-1992
(Permanence of Paper).

CONTENTS

FOREWORD

David Kaiser

In this fascinating essay, renowned physicist and Nobel laureate Alain Aspect describes not one but two quantum revolutions. The first revolution unfolded during the opening quarter of the twentieth century, centered on the efforts of Max Planck, Albert Einstein, Niels Bohr, Werner Heisenberg, Erwin Schödinger, Paul Dirac, and a tight circle of their colleagues. Nearly all contributors to the first quantum revolution understood their own work in revolutionary terms—they recognized, in real time, that a momentous shift was under way in how scientists thought about the world.

The second quantum revolution began as

a spark in the mid-1960s and grew during the 1970s and 1980s. But it only became recognizable as a revolution with hindsight, during the 1990s and early 2000s. In fact, as Aspect describes, researchers around the world are still working hard to realize the full impact of this second revolution.

Seeds for the second quantum revolution were sown during the first. Several legendary architects of quantum theory, including Einstein and Schrödinger, identified the concept of quantum entanglement during the 1930s. They remained skeptical that nature could really display what Einstein famously dismissed as "spooky actions at a distance."

In 1964, John Bell derived a now famous inequality that sets a strict upper limit for how strongly correlated the outcomes of measurements on two or more particles can be, if nature behaved according to the criteria that Einstein and Schrödinger assumed. In particular, Einstein and Schrödinger (and, indeed, Bell him-

self) believed that the outcome of a measurement performed on a particle at one location
should not depend on actions undertaken at an
arbitrarily distant location—a framework often
referred to as "local realism."

Bell's theoretical work clarified that
quantum mechanics is not compatible with
local realism. As he demonstrated, quantum
mechanics predicts that measurements on pairs
of particles in entangled states can be *more
strongly correlated* than the upper limit set by
Bell's inequality—their behaviors somehow
connected in just the way that Einstein had dismissed as too spooky.

After Bell's work, the clash between
quantum theory and local-realist alternatives
seemed clear enough on paper. Left open was
whether nature behaved in accordance with
the quantum-mechanical description, "spooky
actions" and all.

By the early 1970s, a handful of physicists had set out to test Bell's inequality with

new types of experiments, to try to measure whether entanglement really occurred
in the world. Their results were mostly
encouraging—most efforts yielded results
consistent with quantum theory. But none was
definitive.

Aspect and his colleagues designed and
conducted a remarkable experiment. They
inserted fast-changing switches into the paths
that their quantum particles would take, en
route to their respective detectors. Depending
on the orientation that a given switch happened to be in when a particle arrived, that
particle would be directed toward one of two
measuring devices, primed to measure distinct
properties of the particle.

The timing involved was exquisite. The
blink of a human eye lasts about one-tenth of
a second. The switches in Aspect's experiment
changed *ten million times* faster. The quick
changes were required so that the path a given
particle would follow, toward one type of mea-

surement or the other, would change at least one time while each pair of particles was in flight. In addition, the measurements on each member of a pair occurred close enough in time, yet far enough apart in space—separated by several meters across a large laboratory room—that no information traveling at the speed of light could have informed the detector on one side of the room about relevant activities on the other side.

Even with this remarkable arrangement, Aspect and his group measured a clear violation of Bell's inequality, consistent with the predictions of quantum theory. That is, their measurements clearly revealed the uncanny, strong correlations between the behaviors of entangled particles, the possibility of which had so disturbed Einstein decades earlier. Moreover, the odds that the results from Aspect's experiment could have been a statistical fluke, and that nature actually behaved as Einstein had insisted, fell to less than one in a million.

Aspect's beautiful experiment, completed in 1982, had a catalyzing effect on the scientific community. Following his experiment, annual citations in the scientific literature to Bell's work, which had previously languished without much notice, doubled. The tiny spark of the second quantum revolution began to grow.

Remarkably, the achievements by Aspect and other pioneers of this second quantum revolution—including John Clauser and Anton Zeilinger, with whom Aspect shared the Nobel Prize for Physics in 2022—now stretch well beyond deep questions about our most fundamental understanding of nature. A few years ago, the *Economist* magazine reported that research expenditures around the world in the field of quantum information science had already surpassed one billion US dollars per year. The flourishing field of quantum information science—the fruits of the second quantum revolution—builds upon phenomena like quantum entanglement for such next-generation technologies as quantum encryption, quantum computing, and quantum sensing. Research in these areas now spans academic laboratories, government facilities, and private industrial efforts around the world.

How far indeed we have come from Einstein's late-night musings about "spooky actions," or Bell's elegant inequality. Time will tell what subtle mysteries might yet be uncovered during this second quantum revolution—perhaps sparking a third!

EINSTEIN AND THE QUANTUM REVOLUTIONS

Two Quantum Revolutions

Many scientists say that the twentieth century witnessed two great revolutions in physics: relativity and quantum physics.

Both revolutions challenged dramatically the established image we had of the world. But quantum physics radically transformed our lives, and we may speak of a "quantum revolution" just as we might invoke the nineteenth century's "industrial revolution."

The quantum revolution not only disrupted established theories in physics, but also changed the society in which we live. The understanding of the basic laws of thermodynamics and the invention of the steam engine provoked the industrial revolution. Similarly, without a deep understanding of the quantum world, we would not have been able to invent computers or the lasers that enable the processing and transmission of information. No one tinkering in a garage, in California or elsewhere, would have been able to invent the integrated circuits that are the foundations of computers. The information and communication society we know today would not exist without the quantum revolution.

In fact, when we look a bit more closely at the history of quantum physics during the twentieth century, we discover that there were not just one, but two quantum revolutions. The first began early in the century with Max Planck, then with Albert Einstein and his corpuscular description of light, and, in 1923, with Louis de Broglie's undulatory description of particles.

Then, around the 1960s, just as new insights from the quantum perspective seemed to wear thin, scientists realized the importance of entanglement, a concept identified as early as 1935 by Einstein, then by Erwin Schrödinger. Beginning in the 1970s, progress in experimentation made it possible to perform much more intricate experiments, and to observe directly the most extraordinary properties of pairs of entangled objects. These advances led to the second quantum revolution, one that is currently taking place before our eyes.

Let's now take a closer look at these two quantum revolutions.

The First Quantum Revolution

The first quantum revolution occurred in the early twentieth century, when Max Planck studied the radiation emitted by a heated body. Specifically, Planck investigated the radiation emitted by what's called a "blackbody"—a theoretical type of matter that absorbs all radiation, including visible light, and thus would appear completely black. Such a blackbody also emits radiation when heated, and physicists realized that calculating the properties of that "blackbody radiation" as a function of the temperature was a fundamental theoretical problem. To find a solution that agreed with the experimental results, Planck was forced to admit that hot bodies can exchange energy with radiation only in discrete packets of finite size. To emit less than this discrete quantity was a bit like trying to pay less than a penny for something in a store. Only a certain minimum amount can be transferred.

In 1905, Einstein proposed an even more revolutionary idea. Radiation itself is made up of discrete particles of energy. These particles would later be named "photons." Einstein's intuition came from looking for an explanation of the properties of what is called the "photo-electric effect." An experimenter shining light on a piece of metal could create an electrical current—but only if the frequency (or energy) of the light was high enough. When the experimenter shone lower frequencies of light on the metal, there was no current.

Imagine a piece of metal with many electrons trapped under its surface. If you provide enough energy, in the form of light, you can free the electrons. Classical physics says that if you just keep providing light, of any energy, eventually you will provide enough total energy to free the electrons. Experimentation showed this was not true. Only light with a high enough frequency would free electrons. This could be true, Einstein theorized, if the light was made of discrete particles whose individual energy was proportional to the frequency, as Planck proposed in his theory of blackbody radiation. Each individual light particle would have to be energetic enough to get the electrons to escape. Einstein immediately deduced precise laws, laws so shocking for classical physicists that no one believed them at the time. Some ten years later, the famous American scientist, Robert Millikan, undertook experiments aiming to demonstrate, as he would later admit, the falsity of Einstein's predictions. After long, impressive, and arduous experiments, he concluded that in fact Einstein was right.

The Nobel committee understood the revolutionary nature of Einstein's hypothesis, and in 1921 awarded him the prize for his theory of photoelectric effect. Einstein did not win the Nobel Prize for his more famous theory of relativity, contrary to what many believe!

According to Einstein, electromagnetic radiation, including light, is made up of elementary packets of energy, or particles, called photons. One reason this argument was so surprising was that it said the exact opposite of what physicists had shown in the nineteenth century. Brilliant physicists, Thomas Young in England and Augustin Fresnel in France, had demonstrated irrefutably that many properties of light can only be explained if light is not a stream of particles, as theorized by Newton, but a wave. In particular, in one famous experiment, when light travels through two slits, it creates interference patterns—bright and dark zones —that can only be interpreted as if the light waves formed ripples that reinforced each other or canceled each other completely. With the discovery by Maxwell of the possibility of electromagnetic waves traveling at the speed of light—300,000 kilometers per second (186,000 miles per second)—the picture was complete: light is an electromagnetic wave.

But a particle is localized in space, while a wave is widely extended. So how could the two views be reconciled? In 1909 Einstein proposed the now famous wave-particle duality: light was both a particle and a wave. In 1923 Louis de Broglie proposed that this dual nature must also apply to matter. Whether photons (radiation) or electrons (matter), according to these physicists, wave-particle duality should apply. In 1925, two completely different mathematical formalisms were discovered, which expressed this wave-particle duality. Their connection was at first difficult to understand. On the one hand, Erwin Schrödinger developed an equation for the total energy of a system. His "wave equation" governed what is called the wave function, describing matter waves. At the same time, Werner Heisenberg developed what's

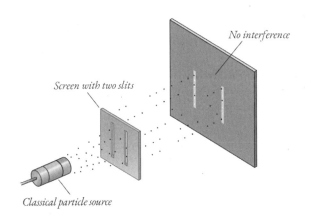

No interference

Screen with two slits

Classical particle source

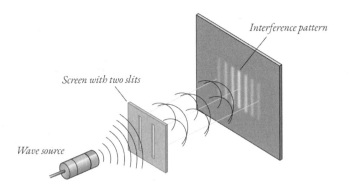

Interference pattern

Screen with two slits

Wave source

called matrix mechanics, which accounted for all the properties of quantum particles. It took a few months for Schrödinger to understand the equivalence of these two formalisms. It was not until the early 1930s that a unifying mathematical formalism, devised by Paul Dirac, was able to account simultaneously for both the wave and particle properties of light, as well as, in a symmetrical way, the wave properties of material particles, for example, electrons in matter.

(*a*) (*facing, top*) In a classical particle model, each particle follows a trajectory through one slit only and lands on the corresponding band on the screen. (*b*) (*facing, bottom*) In a classical wave model, the wave passes through the two slits simultaneously, resulting in an interference pattern on the screen, as observed in the early nineteenth century in Thomas Young's and Augustin Fresnel's experiments on light. Based on illustrations by Ümit Kaya via LibreTexts.

Wave-Particle Duality

Though the mathematical formalism that made it possible to describe wave-particle duality was perfectly coherent, its interpretation was controversial. Most famously, there was an extraordinary clash of titans between Einstein and Niels Bohr, which unfolded in the period from 1925 to 1930.

Einstein can be considered one of the most important contributors to the advent of quantum physics. He not only discovered several remarkable quantum phenomena, but he also realized that there was a major problem in articulating the wave and the particle characteristics of light. At the same time, he would draw major conclusions from these new ideas by establishing the laws that describe the emission and absorption of light by atoms. These are laws that we still use, for example, to describe how lasers function. But, when the new mathematical formalism rendering an account of the wave-particle duality was developed in 1925, it was linked to a probabilistic description of the world, which Einstein did not find at all satisfactory. More precisely, for a given situation, the quantum mathematical formalism does not predict with certainty which of only two possible results will occur: it simply enables the calculation of the probability of obtaining one or the other.

For Einstein, a fundamental physics theory had to predict a precise outcome for every situation. He was therefore not satisfied with the existence of Heisenberg's Uncertainty Principle—which I prefer to call the "dispersion relation"—as it says that it is impossible to know simultaneously with great precision a certain number of physical properties. For example, if an electron is in a quantum state in which its velocity has a quite well defined value, with a small spread, then its position may have any value in a wide interval. Einstein thought this revealed the limits of quantum theory, and that there had to be a more profound theory describing the exact values of both velocity and position of a particle.

He then proceeded to imagine "thought experiments"—the famous *Gedankenexperiment*. At the time he imagined these experiments, they were impossible to perform, given limitations in experimental apparatuses, but were feasible in principle. Thought experiments had to obey all the known laws of physics, even if, in practice, the experiments could not be carried out in a laboratory. Einstein had formerly used thought experiments to discover the laws of relativity. He imagined the latter one in hopes they might show that the limitations imposed by Heisenberg's dispersion relations

The 1927 Solvay Congress in Brussels. Einstein appears in the center of the photograph. Bohr appears on the far right, second row from the bottom. Marie Curie, Werner Heisenberg, Paul Dirac, Louis de Broglie, Erwin Schrödinger, and many other famous physicists are also pictured. Courtesy of ETH-Bibliothek Zurich, Image Archive.

were evidence of an incomplete theory. A more complete theory, he hoped, remained to be discovered.

No doubt, wave-particle duality is extraordinary, and Einstein's unease is understandable. The epic discussions between Einstein and

Bohr, particularly at the 1927 Solvay Congress, are still famous. Einstein imagined the strangest possible situations to demonstrate that quantum theory was incomplete, and for each Bohr offered a counterargument. Bohr showed that all physical laws—including even those discovered by Einstein himself, in particular, general relativity—did not conflict with quantum theory.

I was able to acknowledge this duality directly, during an experiment carried out in the early 1980s with my doctoral student Philippe Grangier. We built a source emitting well-separated single photons. It was the first time such a source had been developed. We then sent photons one by one onto a "beam splitter." What is a beam splitter? Picture a piece of glass onto which falls a ray of sunlight or a laser beam. A portion of the light hitting the glass transmits through, another portion reflects. There are "50–50" beam splitters that allow 50 percent of the light beam to pass through while reflecting the remaining 50 percent.

If we send a photon onto a beam splitter, and if we seriously consider that photon to be a particle, will it split in two? No. An elementary particle cannot divide. So, it will go either one way or the other. Using sophisticated photon-counting systems, Grangier and I demonstrated that this is indeed the case. The photon was found either in one or in the other output channels of the beam splitter, never in both simultaneously. A priori, this is nothing extraordinary for a particle.

But what is astonishing appears when we remember that light is not only made up of particles, but must also be considered a wave. Now, if a wave strikes a beam splitter, what happens? The wave splits in two, and we get two output beams. One can prove this experimentally by using two mirrors to recombine the two beams leaving the beam splitter and observing interference fringes—that is, those bright zones and dark zones from the nineteenth-century light experiments discussed earlier. When the two secondary waves recombine, their oscillations can either be in phase (and join together to create large oscillations) or out of phase (and cancel each other out). The points where they join form bright zones, which physicists call bright fringes; those where they neutralize each other, dark fringes. Thus, if light really is a wave, we should see these interferences. This was the reasoning of Young and Fresnel. So we used the same test with single photons, one by one, and we observed an interference pattern, a smoking gun for a wave-like behavior.

But how can we observe bright bands and dark bands with a single photon? A single photon yields only one bright spot. The experiment has to be repeated many times. Statistically, we then notice that the photons almost always arrive at the locations where bright fringes should appear in a typical interference pattern and never on the dark fringes. After accumu-

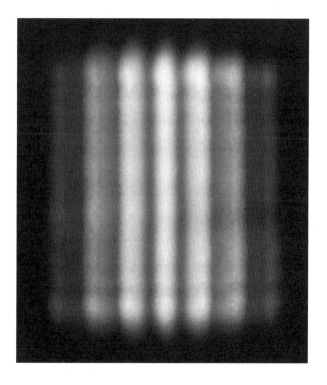

lating a large enough number of photons, no doubt was left: the single photon indeed behaved like a wave since it landed almost always at the place where a usual light wave would show a bright fringe.

The interference pattern created by a light beam passing through two slits (*facing*) is the same pattern produced by a buildup of electrons passing, one at a time, through two slits (*right*). Eleven electrons have passed through the slits in (*a*), 200 in (*b*), 6,000 in (*c*), 40,000 in (*d*), and 140,000 in (*e*). Single photons create a similar interference pattern when sent, one at a time, through two slits. Alain Aspect and Philippe Grangier developed a photon source to show this. Courtesy of Aleksandr Berdnikov (*facing*) and Akira Tonomura (*right*).

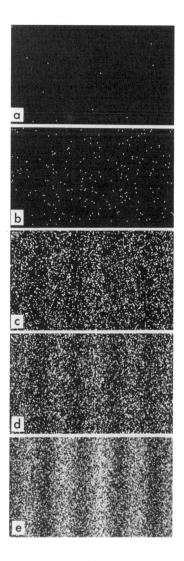

Back to the interference experiment. The photon wave splits in two on the beam splitter and recombines. It thus travels on both sides simultaneously. But the first experiment showed that when the photon arrived on the beam splitter, it went either to one side or to the other, but not to both sides at the same time. How can we reconcile these two observations? This is the conundrum.

According to Bohr, it is complementarity that makes it possible to overcome the epistemological difficulty. What does this mean? It is not possible to conduct both experiments simultaneously. You have to choose: if you put two detectors behind the beam splitter, one in the transmitted path and the other in the reflected path, you will find that the photon goes either to one side or to the other, but there is no longer any possibility of interference. If you want to observe the interference, you must remove the detectors and allow the split wave to recombine. Then it is no longer possible to

prove that the particle went to one side or the
other. Thus, depending on whether one wants
to observe a property or the complementary
property, it is necessary to use different and
incompatible apparatuses.

Bohr went even a little further epistemolog-
ically by saying that it is the measuring device
itself that in a way determines the properties
of the objects being observed. If a device can
reveal a particle nature, the photon behaves like
a particle. If the device can reveal wave behavior, the photon behaves like a wave. Stated
in this way, however, this point of view is too
naïve, as we learned from the eminent physicist
John Archibald Wheeler. Wheeler died in 2008,
just as we were performing the thought experiment he had imagined, known as "the Wheeler
delayed choice experiment." Wheeler had
taken seriously Bohr's idea that it is the measuring device that determines the properties of an
object. He imagined that the photon, when it
strikes the beam splitter, "notices," for instance,

that the device is intended to determine which way it is going, and then behaves like a particle. But if the second device is used, when it strikes the beam splitter, the photon "notices" that; this time, it is intended to reveal a wave-like behavior, and it then behaves accordingly. It is a colorful way of conveying the fact that the device determines the property that is ultimately observed.

Wheeler's brilliant idea was as follows: nothing forces us to choose between the two apparatuses at the very moment when the photon strikes the beam splitter. If the apparatus is large enough, one may delay the moment when the device after the beam splitter is put in place, so that, when the photon strikes the splitter, it "does not know" whether it should go one way or the other, or divide in half to go both ways at once. This is Wheeler's delayed choice experiment.

Since our 1986 experiments, both Grangier and I had been convinced of the importance of Wheeler's thought experiment, but the technological means did not exist for us to perform it at that time. Two decades later, the situation had changed, and, at our suggestion, this experiment was carried out in 2007 at the Institut d'Optique d'Orsay by Jean-François Roch and his students from the École Normale Supérieure de Cachan (now ENS Paris-Saclay). In the 2007 experiment, an interferometer about 50 meters (160 feet) long allowed one to choose to detect either the wave or the particle nature 20 nanoseconds after the photon passed the first beam splitter, when the photon was in the middle of its journey between the first beam splitter and the detecting devices. The experiment showed that one indeed observes the nature corresponding to the final device; when the photon crosses the first beam splitter, the configuration has not yet been chosen. This indicates just how difficult it is to conceptualize wave-particle duality.

But I like to emphasize with my students

that when it comes to doing calculations, we have a single mathematical formalism that describes perfectly well both experiments, and even the delayed choice experiment. We do not have to choose. The formalism of quantum optics has in it both the wave-like and the particle-like behaviors; we can therefore be confident in this theory, which has proven extraordinarily successful.

The Success of the First Quantum Revolution

Quantum theory, as it was developed between 1925 and 1930 by Schrödinger, Heisenberg, and Dirac, is a coherent mathematical formalism that made it possible to understand previously incomprehensible properties of matter. For example, we have known since the end of the nineteenth century that matter is made up of positive and negative charges. And we know that positive and negative charges attract each other. So why does matter not collapse on itself? Classical physics could not provide an answer. Without quantum physics and its wave-particle duality, we cannot understand why an electron, which revolves around a nucleus, doesn't eventually crash into it, just like satellites orbiting around the Earth, which eventually fall into our atmosphere.

How do quantum physics and wave-particle duality make it possible to understand the stability of matter? It is relatively simple if we imagine bringing the electron closer and closer to the atom's nucleus, that is, to confine it within a smaller and smaller amount of space. If we remember that we are dealing with a wave, we must imagine the wave's frequency increasing.

To understand this, remember that if you shorten the length of an organ pipe, the sound becomes higher pitched. The same is true of matter-waves. If we attempt to confine the electron in a smaller and smaller area, the frequency associated with its matter wave will be higher and higher, and, as we know from Planck and Einstein's formula, the energy of the electron becomes greater and greater. For an electron to crash into the nucleus, it should have more energy than the system can give to

The ATLAS particle detector, part of CERN's Large Hadron Collider. The presence of a worker (*center, bottom*) indicates the scale. ATLAS Experiment © 2022 CERN.

it. In other words, there is a limit to how close the electron can get to the nucleus, and this is what explains the stability of matter.

To better understand this counterintuitive reasoning, that the electron demands more energy to get closer to the nucleus, we need only think of the large particle accelerators. Picture the famous one at CERN, near Geneva, Switzerland, for example. Why are particles accelerated? Why are they given ever greater

energy? To probe matter at always smaller
scales. At CERN, we know where this energy
comes from. We use a staggering amount of
electrical power to speed up the electrons that
will then probe matter. But no such outside
source of energy is available for the election in
the atom. The electron cannot therefore come
too close to the nucleus.

Thus, quantum theory, based on a mathematical description that incorporates wave-
particle duality, has made it possible to understand the stability of matter; the workings of
chemical bonds; the mechanical, electrical, and
thermal properties of solid bodies; how electrical current is conducted; how insulators work,
and more. Quantum theory has allowed us to
describe accurately and in detail the exchanges
between matter and light, that is, the emission
and absorption of radiation. It also allows us to
understand even more extraordinary phenomena, such as the superconductivity of certain
metals, or the superfluidity of liquid helium at

extremely low temperatures. Quantum theory is therefore very effective for understanding many physical phenomena, from the simplest to the most complicated.

But there is more! Out of this deep understanding of matter and light, researchers have invented two devices essential to modern life. The first is the transistor. This was the brainchild of the greatest physicists of the 1940s, seeking to better understand the ways in which electrical current could pass through strange materials called semiconductors, the foundation of our computers. Engineers had to develop an extraordinary purification technology for semiconductors, which made it possible to manufacture transistors. Some decades later, they understood how to pack many transistors on a chip, to create integrated circuits, the heart of computers.

The other child of the first quantum revolution is the laser, allowing high-speed communication through optical fibers. Only physicists who fully understood, in a quantum context and with the conceptual difficulties of wave-particle duality, how matter absorbs and emits light could have created lasers.

The Second Quantum Revolution

We have just discussed the first quantum revolution. Then, in the 1960s, scientists came upon a surprising—for lack of a better word—concept beyond wave-particle duality. Einstein and then Schrödinger had been the first to acknowledge it. In 1935, Einstein and his collaborators, Boris Podolsky and Nathan Rosen, wrote a paper that is still famous today, in which they explained that the formalism of quantum mechanics makes it possible to imagine a pair of particles in a strange "entangled" state. (The term "entanglement" itself was coined by Schrödinger, who was pondering the same problem at the same time.)

What is strange are the abnormally strong correlations between those two particles, even when they are very far apart. To explain what this involved, we will consider a scenario imagined by David Bohm, expanding on the reasoning of Einstein and his colleagues. The Bohm thought experiment is based on physical quantities, for example, the polarization of a photon, the measurement of which has only two possible outcomes. In Bohm's version of the situation described by Einstein, Podolsky, and Rosen, each of the two polarization measurement results can occur with equal probability, and the outcome is random for any one particle. It's the same as tossing a coin. But if we compare the results for two *entangled* particles, we find that they are identical each time. If one particle is found with a given result, the other one is found with the same result. The random results are said to be completely correlated. It is as if two magic coins always gave the same result—heads or tails—although tossed by two players separated far from each other. Two coin-toss players would obtain the same results for tosses at the same time, while each of them nevertheless would have the impression that their result was random.

Given this prediction, Einstein concluded that he was right when he said that quantum formalism needed to be supplemented. In his opinion, if two objects, which have interacted in the past but are now separated, present a perfect correlation, they must carry within them a set of properties determined in concert before their separation, and which then have survived in each of the objects. This is the case with homozygous twins who share the same chromosomes. Assume they live apart in distant countries. Imagine that, through bad luck, they have a gene that determines a genetic disease that doesn't present until the age of thirty. It will manifest in both of them at the same time, even if one is in Australia and the other in America. Medical doctors observing such correlations will conclude that the disease is due to a common gene, even if they have not yet identified the gene that is responsible.

In similar fashion, Einstein believed that to understand the very strong correlation he had identified in entangled particles, the particles had to have additional properties that quantum formalism doesn't take into account, but which determine the results. This would therefore require supplementing quantum formalism, which would not be an ultimate description of the world.

Bohr responded that this was impossible. He presented several arguments that convinced many physicists impressed by Bohr's justifiable prestige. But, if one takes the time to study Bohr's paper in detail, as I have done, one may find his response not nearly as convincing as his previous refutations about the Heisenberg relations. Nevertheless, in 1935, the argument went no further, because a whole generation of young physicists applied quantum physics with fantastic success. Deep discussions of quantum theory's foundations and interpretation gave way to the wonder aroused by its formidable experimental successes. Bohr's epistemology then became dominant because most of the pioneers trained with his entourage in Copenhagen, and they were the ones who later taught quantum physics. There was a pervasive sense that Bohr's responses, based on the concept of complementarity, to all of Einstein's objections were quite satisfactory, and that there was no need to waste more time on these questions. One can easily imagine how exciting it was at that time to use quantum mechanics to describe newly discovered experimental facts, so that the question of completing the theory was not a priority. Moreover, at that time, the discussion was only epistemological, about the *interpretation* of the quantum formalism, which Einstein accepted and used as well as Bohr. So why bother about a discussion without any practical consequence?

This was the state of affairs until 1964, when John Stewart Bell, a theoretical physicist at CERN whose daily work was to use quantum mechanics to try to understand the properties of elementary particles, reflected on the principles of this quantum mechanics. He used its formalism like everyone else, but he was very uncomfortable with its basic concepts.

He read the paper by Einstein, Podolsky, and Rosen very carefully, and discovered something new when elaborating on it. Take Einstein's position seriously: let's consider the idea that each entangled particle has hidden properties (unknown to quantum formalism) that will determine the measurement results. You will then see, through mathematical reasoning, that the correlations predicted cannot exceed a certain level. That is, there is an upper limit of correlations fixed by what today is called "a Bell inequality."

But then, considering now the standard quantum formalism, Bell found that quantum theory sometimes predicted correlations greater than that limit. The choice between Einstein's and Bohr's positions was no longer simply epistemological. An experiment that showed correlations beyond that limit would refute Einstein's position.

To me, this scenario seems unprecedented in the history of ideas: a philosophical debate on the nature of the world and the notion of physical reality would be resolved through a physics experiment.

Entanglement Measurement Experiments

Bell's paper showed, for the first time, that an experimenter would arrive at different measurable conclusions depending on whether Bohr's or Einstein's idea was correct. But the paper initially went completely unnoticed. A general wave of amazement at the results of quantum physics had rushed over researchers who used these results without overly considering the fundamental debates between Bohr and Einstein, both of whom had died in the meantime. Theirs was in a certain sense a debate from another time. But it was also partly John Bell's own fault, since he had published his paper in a short-lived journal, one that would have only four issues. Today, you can find the paper easily on the internet, but in 1964 almost no one had access to the paper. Even in 1974, when I became aware of the subject, there were only a handful of original copies of it in a few libraries around the world. I was fortunate to receive a photocopy from Christian Imbert, who became my thesis advisor. He had obtained it from Bernard d'Espagnat, who had himself received it from the American physicist Abner Shimony. It would take several years for physicists to realize the importance of the 1964 paper, and to ask the question: can one realistically conduct the experiment that seems to follow from Bell's theory? Between theory and experiment there is always a big step.

Four physicists then intervened: John Clauser, Michael Horne, Abner Shimony, and Richard Holt. In 1969, they showed how John Bell's thought experiment could be carried out with photons. If one excites an atom on a certain atomic level, it should re-emit two entangled photons in a state that resembles those envisaged by Bell, an Einstein-Podolsky-Rosen-Bohm state with photons. If one makes polarization measurements on these photons, one should be able to reveal the correlations and know whether or not they exceed the Bell limit.

Classically, polarization is the direction of vibration of the electrical field of a light wave, in a plane perpendicular to the direction of propagation. Therefore it can assume any orientation on this plane. To measure it, we use a polarizer, which has an axis such that a polarized light along this axis is transmitted, while light polarized perpendicularly is reflected. If the light is polarized in any direction, a fraction will be transmitted and another reflected.

Once again, a single photon cannot split in two, but will exit in one of two output channels, the transmitted or the reflected one. We therefore perform a polarization measurement that can give only two possible results, which we will

call +1 or –1. When we study a pair of entangled
photons, we make a measurement on the first
and a measurement on the second. By repeating
the experiment many times, we can first check
that each result appears random. But we can also
obtain the probabilities of joint detection, for
example, +1 for the first photon and –1 for the
other photon of the same pair, or any of the four
possibilities. This makes it possible to determine
the level of correlation—that is, whether the
two measurements are independent or connected. The results of the measurement depend
on the orientations of the two polarizers. We
can therefore determine how the correlation
evolves in function of their relative orientations.
It is this set of measurements that would be
subjected to the Bell inequalities test. Either the
limit would be exceeded, or it would not.

These are very tricky experiments. The first
two took place between 1972 and 1973. In fact,
they were nowhere near as simple and clear
as just described. For example, at the time,

polarizers gave only one of the two possible results. The result that corresponded to the other polarization did not exist, the photon was absorbed rather than reflected. One had to do fairly extensive data analysis to say that, if the photon had not been observed, it was probably because it had gone into the complementary, undetected channel.

Two experiments, however, were conducted, one at Harvard University, the other at the University of California, Berkeley, with contradictory results. At Harvard, it was found that there was no violation of the Bell inequalities. But at Berkeley, Clauser and Freedman found that the results agreed with quantum mechanics and did violate the Bell inequalities! The Harvard experiment was repeated sometime later by John Clauser, and then, in a variation, by Ed Fry at Texas A&M. It was agreed that the Berkeley experiment was more convincing than Harvard's, but still with schematics rather far from the ideal experiment.

In the mid-1970s I decided to undertake this type of experiment. When I read Bell's paper, I was enthralled by the clarity of his reasoning, and I was struck by one point in particular: for the test to be fully convincing, it would have been necessary for both ends of the experiment—that is, the two measurements made with the two polarizers—to not be able to exchange signals. Otherwise, one could imagine that an unforeseen interaction, yet to be discovered, allowed the two ends to somehow synchronize their results. In this case, Einstein's perspective again became compatible with quantum correlations. It was therefore necessary to prevent the two ends of the experiment from communicating through any interaction whatsoever.

For this there was an absolute weapon—relativity—which sets an ultimate speed limit for all signals. No signal can go faster than the speed of light. So, if I were able to take simultaneous polarization measurements at both ends, with the direction of the polarizers chosen at the last moment—so that no information about the chosen orientation could go from one polarizer to the other, unless at a velocity faster than light—then I would be certain that no signal had allowed the two sides to communicate.

I carried out this experiment in the early 1980s at the Institut d'Optique d'Orsay with the help of two engineers, Gérard Roger and André Villing, and two students, who have since become eminent physicists: Philippe Grangier and Jean Dalibard. We took these polarization measurements at defined moments with an accuracy on the order of a few nanoseconds, at more than a dozen meters (some 40 feet) apart. To travel this distance, it takes light forty nanoseconds. The two sides of the experiment therefore could not communicate with each other.

We observed the violation of the Bell inequalities, which means that the correlation obtained cannot be explained by the fact that the particles carry within them unmeasured properties. Bohr's idea had triumphed over Einstein's. This result is even more amazing than wave-particle duality. There seems to be an instantaneous exchange between two particles that, at the time of measurement, are 12 meters (almost 40 feet) apart.

About fifteen years later, Nicolas Gisin in Geneva and Anton Zeilinger in Innsbruck repeated a similar experiment with two photons sent into optical fibers in opposite directions, which enabled measurements at distances of around a kilometer (more than half a mile) or even greater. They thus confirmed that, regardless of the distance between particles, they behave as an indivisible, inseparable whole, so inseparable that the connection between them seems to defy relativity. This is called "quantum nonlocality." Despite the distance between them, they continue to have properties that are not local, but global. In 2017, a Chinese team led by Jian Wei Pan showed that entanglement is observable at more than 1,000 kilometers (600 miles), using entangled photons emitted from a satellite and received at ground stations.

Physicists had been noticing this entanglement property for a long time, for example, when trying to describe the helium atom. The two electrons that are attracted by the nucleus of this atom are in an entangled state. But it is on a very small scale, at distances on the order of a few nanometers (a nanometer is one-billionth of a meter), that a lot of strange things occur.

Today, we know that these unusual properties survive at macroscopic distances. Rereading Einstein's texts on the subject, one assumes that he would have been extremely surprised by these results, which sometimes lead people to say that they prove Einstein was wrong. I prefer to stress that he first understood the extraordinary nature of entanglement. So, he can be credited, to some extent, with introducing the new rupture that occurred when imaginative physicists like Richard Feynman said to themselves, *Since this is so bizarre, since this is so revolutionary, can't something useful be done with it?* We then witnessed the development of what today is called quantum information, a thriving new field.

The Manipulation of
Quantum Objects

Before moving on to quantum information,
let's look at another facet of the second quantum revolution, one as important as entanglement. Starting in the 1960s, physicists suc-
ceeded in observing individual microscopic
objects, controlling them, isolating them, and
manipulating them. Up to that point, information on microscopic objects was acquired only
through measurements of large ensembles of
such objects. For example, by illuminating the
atoms of a vapor with light, and by analyzing
the re-emitted light, scientists could determine
the properties of the billions of billions of
atoms in the vapor.

In experiments of this type, the statistical, probabilistic predictions of quantum physics are not too disturbing. If microscopic objects could only be observed statistically through large groupings, it is perhaps appropriate that the theory that naturally describes them is a statistical theory. Eminent physicists, like Schrödinger, thought that we would probably never be able to observe a single electron or atom. Using a clever analogy, Schrödinger wrote that if someone claimed to be observing a single electron in the nearby lab, it would be no less shocking than if a colleague said there was a dinosaur living in a nearby zoo. However, beginning in the 1960s, progress in experimental physics linked to technological progress, especially in electronics, made the first shocking possibility a reality.

For example, in the early 1970s, there was Hans Dehmelt, who managed to trap a single electron for weeks and take measurements on it. It was absolutely astounding. At the end of the 1970s, scientists were able to trap an ion, which, like an electron, is a charged particle. By combining electrical and magnetic fields, the ion was confined within a small space. It was even more extraordinary than the case of the electron because, if a single ion is trapped and it is illuminated with a laser beam whose frequency is well chosen, the number of fluores-

A single trapped ion appears in the center of this photograph, taken in 2017. The ability to observe and manipulate single quantum objects (electrons, ions, atoms, photons), starting in the 1970s, was a crucial ingredient of the second quantum revolution. Courtesy of David Nadlinger/University of Oxford.

cence photons re-emitted is so large, from ten to a hundred million per second, that it could practically be seen with the naked eye. As a young student, I remember how extraordinary it was to see a photograph of a single ion fluorescence. Up until then, we had talked about very large groupings, atomic vapor, and all of a sudden, here was a single ion.

We also learned how to produce and observe individual photon pairs. And here, we come back to the question of entanglement, which is observable only between the two particles of the same pair. In a grouping of many entangled pairs, in which one randomly observes a particle of the first type and a particle of the second type, there is practically no chance of finding two twins belonging to the same pair. One cannot then observe the correlation between entangled particles. The success of the experiments conducted in the years between 1972 and 1982 on pairs of entangled photons is linked to the fact that we know how to produce such well-separated pairs in time: we have a first pair, we measure; a second pair, we measure; a third pair, we measure, and so on.

The combination of the two elements—the understanding of the fact that entanglement goes well beyond wave-particle duality, and the possibility of manipulating individual quantum objects—would give rise to profound reflection that would lead to the developments of quantum information and other quantum technologies.

Quantum Information

Quantum information is based on the idea that if we have new physics laws, there must be new ways of transmitting and processing information. In classical information theory, all systems (computers, transmission channels, and so forth) are equivalent. The systems may be quick or slow, powerful or not, but, within a scaling factor, the laws are always the same. Consequently, these laws rule the ultimate limits of the system. If the calculation time to process a given problem, for example, the factorization of a number into prime factors, increases exponentially with the size of the object under consideration, this scaling law will be true regardless of the power of the computer.

If we now bring into play other behaviors, behaviors as extraordinary as entanglement, certain limits disappear. For example, certain calculations, known as "difficult problems," will be tremendously accelerated if one has a quantum computer capable of entangling a large enough number of quantum bits, or "qubits" (pronounced "cue bits").

What is a difficult problem? For a computer scientist or a mathematician working in this field, a difficult problem is one that requires a calculation time that increases exponentially with the size of the object under consideration. For example, internet security uses a coding called RSA, after Ron Rivest, Adi Shamir, and Leonard Adleman, who described the algorithm in 1977. RSA security is based on the impossibility of factoring a sufficiently large number into prime factors. Factoring into prime factors means, for example, breaking down the number 60 into 3 multiplied by 4 multiplied by 5.

Factoring a large number takes considerable time because you have to try all the prime numbers less than the square root of the number.

This calculation difficulty is the foundation of security. A few years ago, on the internet, coding was done on 64-bit numbers. But, with today's more powerful computers, with great effort and networking, we can crack the code. The number of bits is then doubled, going to 128 bits, and we can rest easy for ten years. Going to 256, we enjoy another ten years, and so on. A number of theoretical physicists, applied mathematicians, and theoretical computer scientists have discovered that certain difficult problems, such as the one just described, would become easier to solve if there were a quantum computer. For example, what is called Shor's algorithm, after mathematician Peter Shor, would enable the factorization of a number in a time that would no longer increase exponentially with size, but in a more reasonable way: a polynomial increase.

To implement Shor's algorithm, one would need a perfect quantum computer. What does that entail? It would use quantum bits, qubits, entangled with each other. To give a vague idea of what this means, we must first ask ourselves what an ordinary "bit" is. In a classical computer system, a bit is a value that can be either 1 or 0, but never both. One traditional way to think of the two states is as a light switch, on or off. But we might equally think of the states as a photon which can go either to one side of the beam splitter or to the other.

In the quantum world, a bit can also be 1 or 0. But here is the interesting part: it can also *at the same time* be 1 *and* 0, like the photon that goes to both sides of the beam splitter *at the same time* in the interference experiment. So, a quantum bit or "qubit" is one that can be put in a state that is a superposition of 1 or 0. If we now take two quantum bits and entangle them, we have access to a great wealth of states because we can have 0–0, 0–1, 1–0, and 1–1, and all the combinations of these four possibilities. If we take three qubits, we have eight basic possibilities (0–0–0, 0–0–1, and so on) as well as the combinations of all those possibilities. If we entangle ten quantum bits, we obtain about a thousand basic possibilities. With twenty, we get one million basic possibilities . . . That's exponential growth!

Imagine taking twenty qubits and performing an elementary operation on these twenty entangled quantum bits. This elementary operation will act simultaneously on the million elementary components of this entangled system. This is the basic concept behind the quantum computer, which in principle would offer massive parallelism—that is, in a single operation it would perform a million, a billion, or a million billion operations in parallel. This would break down the exponential complexity barrier. But this possibility presupposes that we have perfect qubits, which is obviously not the case in a real experiment. As in classical computing, where there are also imperfections, one must use error-correcting codes based on redundancy, the fact that the same information is coded several times on different bits: one would obtain the equivalent of a perfect qubit by distributing quantum information over a certain number of imperfect quantum bits. The problem here is the exorbitant cost of this redundant coding: for a single perfect quantum qubit, hundreds of real qubits, perhaps a thousand, would be needed.

At present, we are still quite far from being able to usefully implement a perfect quantum computer, for many reasons. First, even if we had a perfect quantum computer, we wouldn't have quantum software. More precisely, we do not have a systematic algorithm to solve problems using our quantum computer. We have solutions and methods in only a few cases: to factorize, we have Shor's algorithm; for sorting in very large databases, we have something called Grover's algorithm, and there are a few other examples. But there is no systematic method for proposing a solution to any problem.

The second and arguably more serious difficulty is that no one knows how to go about entangling hundreds of thousands of real quantum bits, let alone a million. In 1980, researchers entangled two photons. And it took several decades to entangle a few hundred of them, using extremely sophisticated methods and extraordinary precautions, in the best laboratories in the world. The world record at the beginning of the 2020s did not exceed a few hundred to a few thousand entangled qubits, and these are imperfect, real qubits. The prospect of entangling hundreds of thousands of qubits, probably more, seems remote.

But there appears to be no fundamental physics law preventing entangling so many qubits, and, with that in mind, we can imagine that sooner or later we will get there, either because the technology will have improved sufficiently, or because a new concept will have made it possible to lower the number of qubits needed to correct imperfections. In the end, one hundred perfect entangled qubits would be enough to have the quantum computer of our dreams.

Beyond the prospect of an ideal quantum computer, the quest for the entanglement of a very large number of qubits is a fundamental problem. Indeed, while many physicists believe there is a boundary between the classical world and the quantum world, nothing indicates that such a boundary really exists, or where it is. By attempting to entangle an increasing number of particles, we might discover an insurmountable boundary, and we would have then identified the barrier between the classical world and the quantum world. That would be an extraordinary result.

While waiting for the hypothetical universal quantum computer built on entangled perfect qubits, we have witnessed the development of extremely active research on quantum simula-

tors using imperfect qubits. The basic idea goes
back to Feynman's influential 1982 paper, in
which he explained that no classical computer
would enable a description of a large number
of entangled electrons in a solid. To study such
a system, he then proposed replacing the electrons with objects that are easier to observe.
This is what is done in a certain number of laboratories that use ultra-cold atoms to simulate the
behavior of electrons in a solid. The advantage
of using ultra-cold atoms is that we can control
them on a quantum scale. (For those interested
in the technical details, for example, in my
laboratory, we place ultra-cold atoms in a dis-
ordered light potential to simulate the behavior
of a noncrystalline matter, and in particular the
celebrated Anderson localization phenomenon.
By contrast, other laboratories use sophisticated
periodic light potentials to simulate crystalline
solids, and attempt to understand certain super-
conducting matter.) Some laboratories can trap
hundreds of atoms individually and combine

these various traps to create simulators that respond to specific questions.

Toward the end of the 2010s, researchers made another surprising discovery, one relating to these simulators based on imperfect entangled quantum bits (the simulators are referred to as NISQ, for "noisy intermediate scale quantum simulators"). They made it possible to find the solution to problems of fundamental physics that we are unable to solve with the computers currently available. Now, it seems also possible to use these simulators to solve difficult optimization problems, that is, those requiring a classical computation time that increases exponentially with the size of the problem. The time saved with a quantum simulator would itself be exponential, with significant practical consequences, which seem within reach. Consider a typical example of a difficult problem: the problem of the distribution of electrical current on the scale of a country or a continent. The optimization of an electrical network must be done in real time, as quickly as possible, especially with new, intermittent energy sources, such as wind turbines. This seems to be within reach with a NISQ simulator.

In connection with quantum computing, the notion of quantum advantage has given rise to many, sometimes exaggerated, statements. Moreover, the notion of quantum advantage is often restricted to the idea that, because of its exponential acceleration of calculations, quantum computing will make it possible to solve problems unsolvable by the most powerful computers today. But we must also consider the possibility that quantum computing can save energy resources, by solving at a much lower energy cost *problems we can already solve* at large computing centers, whose electricity consumption is enormous. This should motivate young people concerned with the problems of the planet. Working in quantum technologies could allow them to combine scientific passion with their environmental convictions.

Quantum Cryptography

Finally, let's talk about a slightly simpler quantum information application, one that represents an already proven success in quantum information: quantum cryptography. It works, and it works so well that some companies are already marketing quantum cryptography systems.

What does this involve? Present internet security, as mentioned above, requires the encoding of information in a cryptographic system, such that people seeking to decode our message will be unable to do so using computers at their disposal. But let's imagine that our adversaries have computers a thousand times more powerful than ours: then these foes can break our codes. This is, for example, what happened during World War II: certain codes used by the German Navy were broken thanks to the first computers running at MIT.

If our adversaries have tremendously more advanced computational means than we do, then these adversaries can break the code we are using. It is also possible that they have a mathematical theorem unknown to us, which enables them to rapidly factorize large numbers. No one has ever demonstrated that such a theorem does not exist, and one cannot exclude its existence. Traditional cryptography is therefore based on the assumption that our adversaries are approximately at our own mathematical and technical level.

Quantum cryptography is of a different nature. In it, security is based on the laws of quantum physics. Whatever the technical level of the people attempting to break our code, the thing that will limit the possibility of their spying on our message is the quantum properties of the code and not our adversary's technical limitations. Why? Because if a spy attempts to decode a photon used to transmit information, he or she will inevitably disturb it, and in such a way that we could notice. So we would immediately stop the transmission before sending any secret data.

The fundamental idea is that, in quantum physics, we cannot make a measurement on an object, or access information about it, without leaving a trace. If we provide ourselves tools to detect the presence of a spy, we will know if there is someone online, or if we can transmit securely.

The two methods of quantum cryptography proposed in the 1980s and 1990s are based on the two basic elements of the second quantum revolution. Charles Bennett and Gilles Brassard's method uses the possibility of controlling and observing individual photons. Artur Ekert's uses pairs of entangled photons.

In the latter case, if we observe a violation of
the Bell inequalities, we can be sure there isn't
a spy online. More recently, these ideas have
been extended to other types of quantum states
of light, known as continuous variables.

Here we have, then, an application that is
obviously of interest to militaries around the
world. But consider an issue equally important
to governments, the social impact of the dis-
closure of confidential information. It is likely
that today most coded information circulating
on the internet is recorded if not decrypted by
adversaries. It is likely that in perhaps a decade,
with more powerful computers, quantum computers, or advances in applied mathematics,
those who have stored this information will
be able to decrypt it. We may then have a new,
very large-scale, WikiLeaks affair in which
information, for example, state secrets powerful
enough to destabilize international relations,
could be made public. To guard against this,
quantum cryptography is a promising solution.

In Search of
the Limit

Quantum physics is extraordinary. After a century of effort, we know of no scenario in which the theory appears to have reached its limits.

This is unusual. In physics, all theories reach a limit. This doesn't mean they're false. Usually, a more general theory will encompass the previous one and supplement it in extreme situations, but the first theory will remain valid within a wide range of conditions. For example, Newton's mechanics remains valid for motion of satellites, or of rockets, but it does not correctly deal with movements at speeds close to that of light. In those cases, one must then use special relativity, which contains Newton's mechanics as an excellent approximation at a lower speed.

Today, we don't even know where quantum physics might reach a limit. In fact, the situation of quantum physics is unusual in that we know perfectly how to write the mathematical formalism to use it, yet we still have difficulty conceptualizing that theory. These entangled particles, which seem to form a whole, even when they are very far from each other, are hard to picture. Physicists manage as best as they can, by developing their own images, but no single picture satisfies everyone. Although everyone agrees on how to use the mathematics, there is much less consensus on providing images of that formalism.

Quantum physics has so far emerged victorious from all the battles that have been waged to find its potential weakness. It has managed to escape all attempts to discredit it. Einstein fought many of these battles, and we should thank him for drawing our attention to quantum theory's extraordinary properties, which continue to offer new possibilities for long-term technological advancements. Again, without wave-particle duality, we would not have today's information society. If quantum information, based on the second quantum revolution, keeps its promises, who knows what transformations society in general will experience? Will we be able to entangle a large number of quantum bits? Will we come up against a fundamental limit? We are in an enthralling chapter in physics—one that will not soon be concluded.

ABOUT THE AUTHOR

A physicist specializing in quantum optics, Alain Aspect experimentally demonstrated in his PhD dissertation, at the Institut d'Optique d'Orsay in 1982, the apparent nonlocality of quantum entanglement. This property, used today in quantum computing and quantum cryptography, reflects the ability of two photons to appear to exchange information, even at a distance from each other. Aspect was a co-recipient of the 2022 Nobel Prize in Physics "for experiments with entangled photons, establishing the violation of Bell inequalities and pioneering quantum information science."

In 1985, with his graduate student Philippe Grangier, he invented the first single photon source and demonstrated the wave-particle quantum duality with such single photons.

In 1987, while a lecturer at the École Polytechnique and deputy director of the Atomic
and Molecular Physics Laboratory at the École
Normale Supérieure and Collège de France,
he worked with Claude Cohen-Tannoudji on
a method for cooling atoms by laser called
"below the one-photon recoil energy," one
of the works that earned Claude Cohen-
Tannoudji the Nobel Prize in Physics in 1997. In
1992, he created the Atom Optics Group at the
Institut d'Optique d'Orsay, devoted to atomic
mirrors, Bose-Einstein condensates, atom
lasers, and ultracold atom quantum simulators.
He is currently professor at the Institut d'Optique (Université Paris-Saclay), professor at the
École Polytechnique (Institut Polytechnique
de Paris), and CNRS senior scientist emeritus.

Alain Aspect received the CNRS Gold
Medal in 2005, the Albert Einstein Medal in
2012, as well as the Niels Bohr Gold Medal
and the Balzan Prize in 2013, and many

other awards. He is a member of the French Académie des Sciences, the French Académie des Technologies, and other science academies in several countries (Austria, Belgium, United States, United Kingdom, and Italy).